PRENTICE HALL
SCIENCE

Activity Book

CELLS
Building Blocks of Life

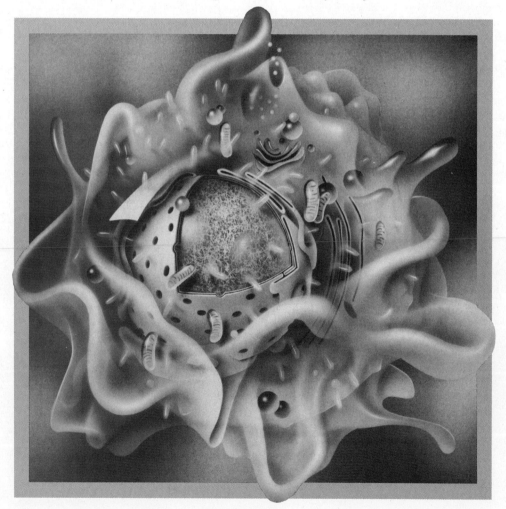

Prentice Hall
Englewood Cliffs, New Jersey
Needham, Massachusetts

Activity Book

PRENTICE HALL SCIENCE
Cells
Building Blocks of Life

ISBN 0-13-400482-5

    8 9 10   97

Prentice Hall
A Division of Simon & Schuster
Englewood Cliffs, New Jersey 07632

Contents

To the Teacher

The materials in the *Activity Book* are designed to assist you in teaching the *Prentice Hall Science* program. These materials will be especially helpful to you in accommodating a wide range of student ability levels. In particular, the activities have been designed to reinforce and extend a variety of science skills and to encourage critical thinking, problem solving, and discovery learning. The highly visual format of many activities heightens student interest and enthusiasm.

All the materials in the *Activity Book* have been developed to facilitate student comprehension of, and interest in, science. Pages intended for student use may be made into overhead transparencies and masters or used as photocopy originals. The reproducible format allows you to have these items easily available in the quantity you need. All appropriate answers to questions and activities are found at the end of each section in a convenient Answer Key.

CHAPTER MATERIALS

In order to stimulate and increase student interest, the *Activity Book* includes a wide variety of activities and worksheets. All the activities and worksheets are correlated to individual chapters in the student textbook.

Table of Contents

Each set of chapter materials begins with a Table of Contents that lists every component for the chapter and the page number on which it begins. The Table of Contents also lists the number of the page on which the Answer Key for the chapter activities and worksheets begins. In addition, the Table of Contents page for each chapter has a shaded bar running along the edge of the page. This shading will enable you to easily spot where a new set of chapter materials begins.

Whenever an activity might be considered a problem-solving or discovery-learning activity, it is so marked on the Contents page. In addition, each activity that can be used for cooperative-learning groups has an asterisk beside it on the Contents page.

First in the chapter materials is a Chapter Discovery. The Chapter Discovery is best used prior to students reading the chapter. It will enable students to discover for themselves some of the scientific concepts discussed within the chapter. Because of their highly visual design, simplicity, and hands-on approach to discovery learning, the Discovery Activities are particularly appropriate for ESL students in a cooperative-learning setting.

Chapter Activities

Chapter activities are especially visual, often asking students to draw conclusions from diagrams, graphs, tables, and other forms of data. Many chapter activities enable the student to employ problem-solving and critical-thinking skills. Others allow the student to utilize a discovery-learning approach to the topics covered in the chapter. In addition, most chapter activities are appropriate for cooperative-learning groups.

Laboratory Investigation Worksheet

Each chapter of the textbook contains a full-page Laboratory Investigation. A Laboratory Investigation worksheet in each set of chapter materials repeats the textbook Laboratory Investigation and provides formatted space for writing observations and conclusions. Students are aided by a formatted worksheet, and teachers can easily evaluate and grade students' results and progress. Answers to the Laboratory Investigation are provided in the Answer Key following the chapter materials, as well as in the Annotated Teacher's Edition of the textbook.

Answer Key

At the end of each set of chapter materials is an Answer Key for all activities and worksheets in the chapter.

SCIENCE READING SKILLS

Each textbook in *Prentice Hall Science* includes a special feature called the Science Gazette. Each gazette contains three articles.

The first article in every Science Gazette—called Adventures in Science— describes a particular discovery, innovation, or field of research of a scientist or group of scientists. Some of the scientists profiled in Adventures in Science are well known; others are not yet famous but have made significant contributions to the world of science. These articles provide students with firsthand knowledge about how scientists work and think, and give some insight into the scientists' personal lives as well.

Issues in Science is the second article in every gazette. This article provides a nonbiased description of a specific area of science in which various members of the scientific community or the population at large hold diverging opinions. Issues in Science articles introduce students to some of the "controversies" raging in science at the present time. While many of these issues are debated strictly in scientific terms, others involve social issues that pertain to science as well.

The third article in every Science Gazette is called Futures in Science. The setting of each Futures in Science article is some 15 to 150 years in the future and describes some of the advances people may encounter as science progresses through the years. However, these articles cannot be considered "science fiction," as they are all extrapolations of current scientific research.

The Science Gazette articles can be powerful motivators in developing an interest in science. However, they have been written with a second purpose in mind. These articles can be used as science readers. As such, they will both reinforce and enrich your students' ability to read scientific material. In order to better assess the science reading skills of your students, this *Activity Book* contains a variety of science reading activities based on the gazette articles. Each gazette article has an activity that can be distributed to students in order to evaluate their science reading skills.

There are a variety of science reading skills included in this *Activity Book*. These skills include Finding the Main Idea, Previewing, Critical Reading, Making Predictions, Outlining, Using Context Clues, and Making Inferences. These basic study skills are essential in understanding the content of all subject matter, and they can be particularly useful in the comprehension of science materials. Mastering such study skills can help students to study, learn new vocabulary terms, and understand information found in their textbooks.

ACTIVITY BANK

A special feature called the Activity Bank ends each textbook in *Prentice Hall Science*. The Activity Bank is a compilation of hands-on activities designed to reinforce and extend the science concepts developed in the textbook. Each activity that appears in the Activity Bank section of the textbook is reproduced here as a worksheet with space for recording observations and conclusions. Also included are additional activities in the form of worksheets. An Answer Key for all the activities is given. The Activity Bank activities provide opportunities to meet the diverse abilities and interests of students; to encourage problem solving, critical thinking, and discovery learning; to involve students more actively in the learning experience; and to address the need for ESL strategies and cooperative learning.

Contents

*Appropriate for cooperative learning

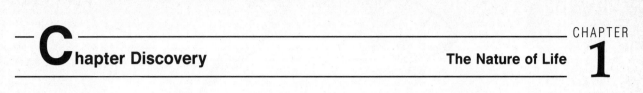

Chapter Discovery

The Nature of Life

CHAPTER
1

Characteristics of Living Things

Background Information

Do you think it's always easy to tell if something is alive? Sometimes it is, and sometimes it's not. It might be easier to decide if you list the characteristics of living things. For example, living things move. You might like to perform this activity again, after you have completed this chapter.

Part A

Look closely at each picture below. Decide whether each item is living or nonliving. If you think an item is living, write "yes" next to the letter. If you think it is nonliving, write "no."

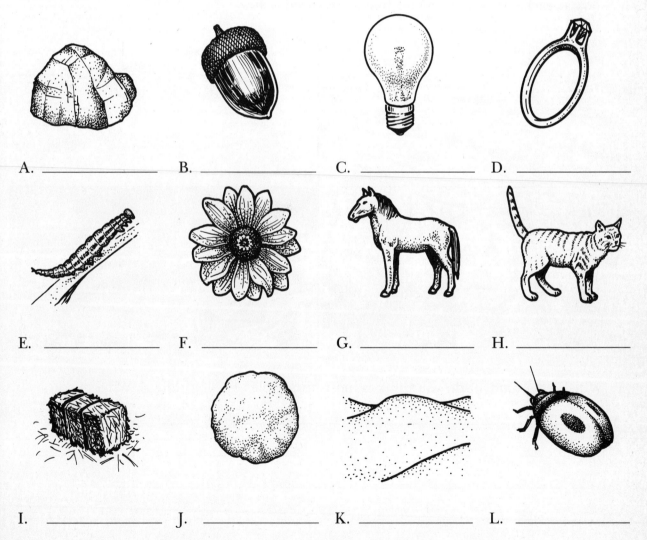

A. _____   B. _____   C. _____   D. _____

E. _____   F. _____   G. _____   H. _____

I. _____   J. _____   K. _____   L. _____

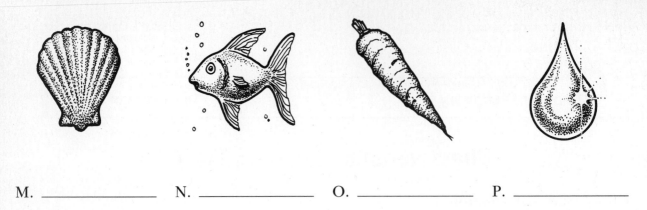

M. _____   N. _____   O. _____   P. _____

Background Information

In nature, living things live in different places. Polar bears live in a cold environment. Lions live in a warm environment. Even though cold and warm environments obviously differ in temperature, there are many other differences as well. For example, plants do not grow well on land in the very coldest environments. Plants grow well in most warm environments, however.

Part B

Look at each of the four environments pictured below.

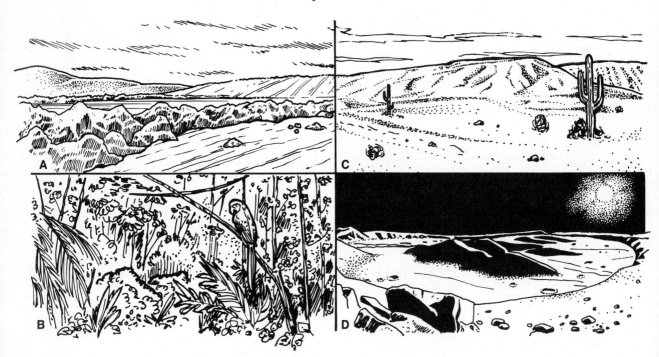

1. Which environment do you think is most hospitable to living things? Why?

_____

_____

_____

_____

_____

2. Which environment do you think is least hospitable? Why? _____

_____

_____

_____

_____

_____

3. Looking at the other two environments, what special problems might living things encounter? What types of living things might be able to survive in these

environments? _____

_____

_____

_____

_____

_____

_____

Critical Thinking and Application

1. Look at the decisions you made in Part A. On what criteria did you base your

choices? _____

_____

_____

_____

_____

2. Were there items in Part A that you were not sure of? Why? _____

_____

_____

_____

_____

3. Based on your answers to Part A, what are some characteristics that all living things seem to share? _____

_____

_____

_____

4. Do you think there are certain substances or "ingredients" that make up all living things? If so, what do you think they are? _____

_____

_____

_____

5. Look at your answers to Part B. What do you think living things need in their environment? _____

_____

_____

_____

_____

_____

**ctivity**

Redi's Experiments

Examine the three sets of drawings below. Each set represents the beginning of one of Redi's experiments. You are to draw in the results of each experiment.

Beginning of the experiment →   **Several days later**

Experiment A

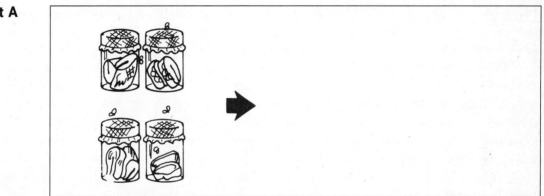

Experiment B

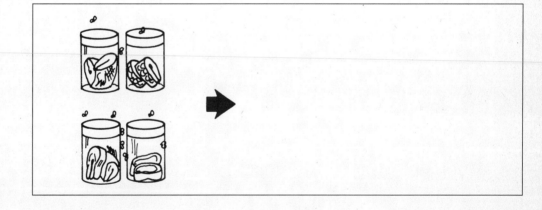

Experiment C

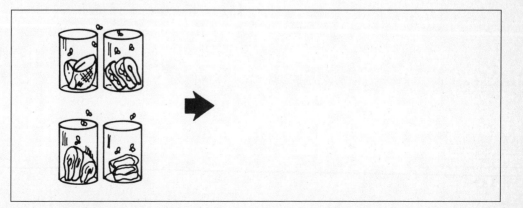

ctivity

Characteristics and Needs of Living Organisms

Using the list below, fill in the chart.

Hops; has four legs
Lays soft eggs on land
Insects, worms, berries, seeds
Deserts, forests
Migrates short distances
Crawls on belly
Queen lays thousands of eggs
Lizards, insects, small snakes
Poisonous; feeds mainly at night
May capture prey by distracting it
Swims using fins
Can change color to match background
Ponds, marshes, streams
Small mollusks, worms, small crabs
Hedgerows, shrubs, farms, and gardens

Lays eggs in water
Can see ultraviolet "color"
Has six legs; walks, flies
Bears live young
Wooded or bushy areas
Pollen and nectar
Lays eggs in water
Flies; walks on two legs
Hibernates in winter
Fields, flower gardens
Lays hard-shelled eggs in a nest
Rats, mice, rabbits, squirrels
Salt water or fresh water
Insects, spiders, earthworms
Walks and runs on four legs

Animal	Movement	Reproduction	Food	Living Space	Response
Bullfrog					
Blackbird					
Flounder					
Honeybee					
Coral Snake					
Red Fox					

Activity

Symbols for Elements

Chemists usually do not write out the name of an element. Instead they use a symbol that is made up of one or two letters. Some symbols, such as O for oxygen, C for carbon, and S for sulfur, are easy to remember. However, some elements have symbols that do not seem to be related to their names. Some of these elements are gold, silver, lead, potassium, tin, iron, and mercury. Use reference books to find out their symbols and how they came to be used for these elements.

Element	Symbol	Reason for Symbol
Gold		
Silver		
Lead		
Potassium		
Tin		
Iron		
Mercury		

_____ *Laboratory Investigation* _____

CHAPTER 1 ■ The Nature of Life

You Are What You Eat

Problem
Does your school lunch menu provide a balanced diet?

Materials *(per group)*
school lunch menu for the current week
pencil
paper
reference book or textbook on nutrition

Procedure
1. Obtain a copy of your school's lunch menu for one week.
2. For each day of the week, place each item from the menu in the appropriate food group in Table 1.
3. Table 2 lists the three major nutrients: carbohydrates, fats, and proteins. List those foods containing large amounts of these nutrients under the proper heading.
4. Identify those foods that are plants or plant products and those that are animals or animal products in Table 3.

Observations
Study the data you have collected and organized.

Table 1

Meat Group	Vegetable-Fruit Group	Milk Group	Bread-Cereal Group

Table 2

Carbohydrates	Fats	Proteins

Table 3

Plants or Plant Products	Animals or Animal Products

Analysis and Conclusions

1. What conclusions can you draw regarding your school's lunch program? _____

_____

_____

2. According to your data, do the foods represent a balanced diet? _____ Do foods in

certain categories appear much more often than foods in other categories? _____

_____

_____

_____

3. What changes, if any, would you make in the menus? _____

_____

_____

_____

4. **On Your Own** Do a similar exercise, but this time analyze the dinners you eat for a week.

Answer Key

Chapter Discovery: Characteristics of Living Things

Part A No: A, C, D, I, J, K, M, P; Yes: B, E, F, G, H, L, N, O **Part B** **1.** Answers may vary, but students will probably select the forest environment because it has reasonable temperatures; sufficient moisture; lots of food, air, and sunlight; and room. **2.** Answers may vary, but most students will select the outer space environment because it lacks air, water, probably has unpleasant temperatures, and has no source of food. **3.** The polar environment is cold and has little sunlight in winter. The desert environment is hot and dry; both seem to have limited sources of food. Animals with warm fur are adapted to cold environments. Plants that live low to the ground and are able to survive cold temperatures and plants and animals that live in the comparatively warm oceans in cold regions can survive. Plants that require little moisture are adapted to life in the desert.

Critical Thinking and Application

1. Accept all reasonable answers. If you choose to ask students to reexamine their choices after they have studied the material presented in this chapter, see if they have changed their minds about any of their answers. See if they are better able to decide after they have completed the chapter.
2. Answers will vary. **3.** Accept all reasonable answers. Some possible answers include: movement, respond to stimuli, grow, ability to reproduce, ability to eat or to make their own food. See if students are better able to answer this question after they have completed the chapter. **4.** Accept all reasonable answers. Elements that are found in all living things include carbon, hydrogen, oxygen, and nitrogen; water is an important compound found in all living things.
5. Possible answers include: light, air or oxygen, water, space to grow, source of food, or a source of the materials needed to make food, room to live, reasonable temperatures.

Activity: Redi's Experiments

Experiment A There should be flies inside and around the open jars and maggots on the meat. **Experiment B** There should be flies around the closed jars and no maggots on the meat. **Experiment C** There should be maggots on top of the netting covering the jars.

Activity: Characteristics and Needs of Living Organisms

Bullfrog: hops, has four legs; lays eggs in water; insects, spiders, earthworms; ponds, marshes, streams; hibernates in winter
Blackbird: flies, walks on two legs; lays hard-shelled eggs in a nest; insects, worms, berries, seeds; hedgerows, shrubs, farms, and gardens; migrates short distances **Flounder:** swims using fins; lays eggs in water; small mollusks, worms, small crabs; salt water or fresh water; can change color to match background **Coral Snake:** crawls on belly; lays soft eggs on land; lizards, insects, small snakes; deserts, forests; poisonous, feeds mainly at night **Red Fox:** walks and runs on four legs; bears live young; rats, mice, rabbits, squirrels; wooded or bushy areas; may capture prey by distracting it

Activity: Symbols for Elements

gold Au aurum **silver** Ag argentum **lead** Pb plumbum **potassium** K kalium **tin** Sn stannum **iron** Fe ferrum **mercury** Hg hydargyrum

Laboratory Investigation: You Are What You Eat

Observations Check each student's data to make sure the data are organized and scientifically correct. **Analysis and Conclusions 1.** In most cases, students will determine that school-lunch programs provide a balanced diet.   **2.** Again, students will usually determine that their school lunch provides a balanced diet. They will probably find, however, that some foods appear on the menus far more frequently than others do.

3. Accept all reasonable answers, assuming that the changes still represent a balanced diet.   **4.** Data and results will vary, depending on the particular student.

Contents

Chapter Discovery

Chapter Activities

Laboratory Investigation Worksheet

   (**Note:** *This investigation is found on page D58 of the student textbook.*)

*Appropriate for cooperative learning

Discovering Cell Function

The cartoon below shows the various functions carried out in a typical plant cell. Study the cartoon; then describe the function that you see being carried out at each number. After you have studied Chapter 2 in your textbook, look at the cartoon again and note the cell structures that carry out each function.

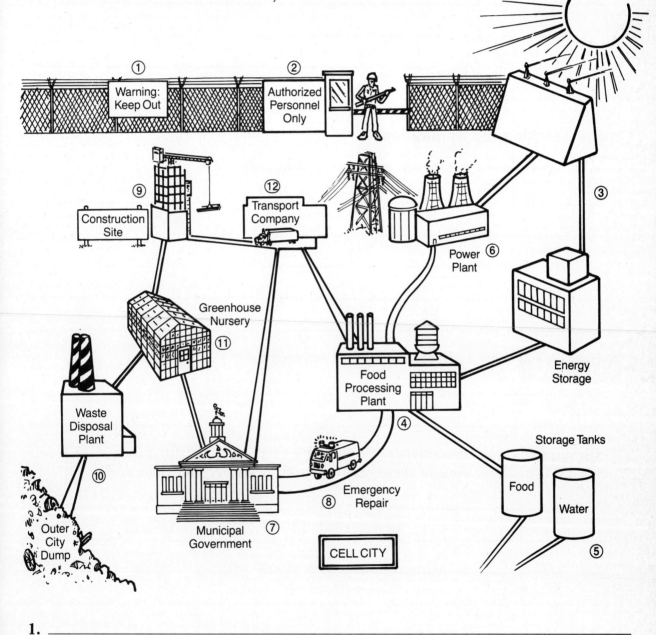

1. _____

_____

2. _____

3. _____

4. _____

5. _____

6. _____

7. _____

8. _____

9. _____

10. _____

11. _____

12. _____

Critical Thinking and Application

1. What activities are carried out in a typical plant cell?

    _____

    _____

    _____

    _____

    _____

2. How do you think a cell accomplishes these activities?

    _____

    _____

3. Do you think that certain types of cells might carry out specialized functions? If so, what do you think some of these cell types and specialized functions are?

    _____

    _____

    _____

    _____

Activity

Can You Identify These Cell Structures?

Read each description and then identify the cell structure. Write your answer on the line provided.

1. I'm a real "powerhouse."
That's plain to see.
I break down food
To release energy.

What am I? _____

2. I'm strong and stiff
Getting through me is tough.
I'm found only in plants,
But I guess that's enough.

What am I? _____

3. My name means "colored bodies,"
And I contain DNA.
I pass on traits to new cells
In a systematic way.

What am I? _____

4. I'm the "brain" of the cell
Or so they say.
I regulate activities
From day to day.

What am I? _____

5. Found only in plant cells,
I'm green as can be.
I make food for the plant
Using the sun's energy.

What am I? _____

6. I'm a series of tubes
Found throughout the cell.
I transport proteins
And other things as well.

What am I? _____

7. I'm full of holes,
Flexible, and thin.
I control what gets out
As well as what comes in.

What am I? _____

8. Proteins are made here
Even though I'm quite small.
You can find me in the cytoplasm
Or attached to E.R.'s wall.

What am I? _____

9. I've been called a "storage tank"
By those with little taste.
I'm a sac filled with water,
Food, enzymes, or waste.

What am I? _____

10. Since I contain many enzymes,
I can digest an injured cell;
And can break down a large molecule
Into a smaller one as well.

What am I? _____

Activity

Shapes of Cells

If your library has a card catalog or an audiovisual catalog, use it to locate a film or filmstrip about the shapes of various cells. If films or filmstrips are unavailable, you may find this topic detailed in a reference book.

Using your sources of information, make drawings of all the different shapes you can find. Take notes about the shape of each cell and describe each cell's function. Detail your findings in a written report.

Activity   Cell Structure and Function

Miracles of Modern Medicine

During the past 200 years, knowledge concerning medical treatment of the human body has grown tremendously. Many of today's—and tomorrow's—medical "miracles" involve the manipulation of cells, tissues, organs, and organ systems.

A time line for the years 1800 to 2000 is presented below along with a list of ten events in medicine. Read each of the events. Then write the letter for each event on the time line. Notice that event A is already on the time line for you. Notice also that one of the events is in the future!

```
                                                          A
 |___|____|____|____|____|____|____|____|____|
1800  1825  1850  1875  1900  1925  1950  1975  2000
```

A. In 1954, the first successful kidney transplant took place.
B. In 1877, Louis Pasteur established beyond any doubt that germs cause infectious disease.
C. In 1978, two British doctors, Patrick Steptoe and Robert Edwards, achieved the live births of two "test-tube babies."
D. In 1967, the first successful heart transplant occurred.
E. In 1900, the use of the X-ray to diagnose lung ailments and bone fractures became a common practice, although X-rays had been discovered only five years earlier. Unfortunately, the harmful side effects of X-rays were not wholly realized.
F. In the year 2000, Jim Clark died of an incurable disease. Before he died, he made arrangements to have his body frozen in liquid nitrogen and kept in a cryogenic capsule. Five years later, when a cure for his disease was discovered, his body was "defrosted" and restored to life.
G. In 1922, Karl Landsteiner discovered the four ABO blood groups and changed blood transfusions from a dangerous experiment to a lifesaving routine.
H. In 1846, John Warren began the age of pain-free surgery with a dramatic demonstration of the use of ether as an anesthetic.
I. In 1981, Swiss researchers developed a clone of gray mice. They did this by inserting nuclei from embryo gray mice into the fertilized eggs of black mice after removing the fertilized nucleus. The "new" eggs were implanted into female white mice who gave birth to gray offspring!
J. In 1819, the stethoscope was invented by René Laënnec.

ctivity   **Cell Structure and Function**

CHAPTER **2**

Biologists Who Studied Cells

Using reference materials found in the library, look up information concerning three biologists: Mathias Schleiden, Robert Hooke, and Theodor Schwann. Find out how each biologist contributed to the study of cells.

In a written report, include the answers to the following questions:

1. When and where did each biologist live?
2. What type of work did each biologist do during his lifetime?
3. What was each biologist's particular contribution to the study of cells?

Mathias Schleiden _____

_____

_____

_____

_____

_____

_____

Robert Hooke _____

_____

_____

_____

_____

_____

Theodor Schwann _____

_____

_____

_____

_____

_____

Activity

Cell Structure and Function

Division of Labor

The levels of organization in an organism represent a division of labor among the various parts so that the work of the living thing gets done.

Your community probably is based on a division of labor also. Examine your community or an aspect of your community. How are the various jobs divided among people? How effective is this division of labor? How does division of labor in your community compare with division of labor in an organism? Write a brief paragraph answering these questions in the space below.

Activity _____ **CHAPTER**

Cell Structure and Function **2**

Shape and Function of Cells

In multicellular organisms, individual cell shape is usually associated with function. To observe this, collect an assortment of plant parts. Take these back to the classroom for further study. Using whatever means of magnification you have available, such as a magnifying glass or a microscope, see if you can determine the shape of the cells from the different plant parts. Draw what you observe in the spaces provided. Label the parts of the cells that you are able to identify. Then answer the questions below.

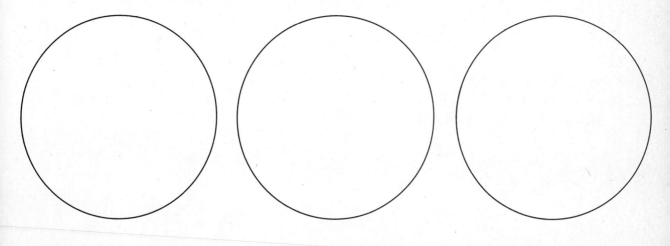

_____   _____   _____

1. How do the different parts of a plant vary in cell shape and size? _____

_____

_____

_____

2. Which cells are cylindrical, spherical, cubical, or spindle-shaped? _____

_____

_____

3. How is each cell's shape related to its function? _____

_____

_____

_____

Activity

Cell Mathematics

Read each of the following statements about cells. On the line provided, either rewrite the underlined numerical expression using numbers, or perform a mathematical computation to fill in the correct answer.

> **Example:** The first cells probably appeared on Earth about 3.5 billion years ago. 3,500,000,000 years

1. The average adult human body consists of about <u>fifty thousand billion</u> cells. _____ cells

2. In a normal adult male, a small drop of blood contains approximately <u>five million</u> red blood cells. _____ cells

3. A drop of blood from an adult female would contain about <u>five hundred thousand</u> fewer cells than that of an adult male. _____ cells

4. The average cell is about <u>eight thousandths</u> of a millimeter in diameter.

   _____ mm

5. A human liver cell has a mass of <u>one millionth</u> of a gram. _____ g

6. Blood is made up of blood cells and plasma. If the blood cells compose 45 percent of the blood, what percentage is plasma? _____ percent

7. A drop of blood contains five million red blood cells and ten thousand white blood cells. How many <u>times</u> more red blood cells are there than white blood cells?

   _____ times

8. A hen's egg is about 50 mm in diameter. A human egg is about 0.14 mm in diameter. What is the difference in size between the hen's egg and the human

   egg? _____ mm

9. The nucleus of a cell is about 0.000625 cm in diameter. What is the diameter of this

nucleus expressed in millimeters? _____ mm

A micron is 0.001 mm. What is the diameter of this nucleus expressed in microns?

_____ microns

10. Convert the lengths shown in microns to millimeters.

cell = 8 microns = _____ mm

bacterium = 0.5 microns = _____ mm

virus = 0.09 microns = _____ mm

11. A single cell reproduces by mitosis. How many cells will be present at the end of the

first division? _____ At the end of the second division? _____

Fifth division? _____ Tenth division? _____

Twentieth division? _____

12. Assume the thickness of a cell to be 0.1 mm. (Although this is very thin, it is still
much thicker than the average cell.) At any rate, assume that each time the cell
doubles, all of the new cells are stacked on top of one another. How many doublings
would it take to produce a stack of cells that would reach to the moon (about

386,232 km away)? _____

13. A biologist notices that a petri dish is becoming covered with colonies of bacteria.
The doubling rate of the bacterial colonies is one day. At the end of 30 days, the
plate is covered with bacteria. When was the petri plate half-covered with bacterial

colonies? _____

_____ *Laboratory Investigation* _____

CHAPTER 2 ■ Cell Structure and Function

Using the Microscope

Problem
How do you use the microscope to observe objects?

Materials *(per group)*
small pieces of newspaper print
    and colorful magazine photographs
microscope slide
medicine dropper
coverslip
microscope

Procedure 🜊

1. Your teacher will instruct you as to the proper use and care of the microscope. Follow all instructions carefully.

2. Obtain a small piece of newspaper print and place it on a clean microscope slide.

3. To make a wet-mount slide, use the medicine dropper to carefully place a drop of water over the newsprint.

4. Carefully lower the coverslip over the newsprint.

5. Place the slide on the stage of the microscope. The newsprint should be facing up and should be in the normal reading position.

6. With the low-power objective in place, focus on a specific letter in the newsprint.

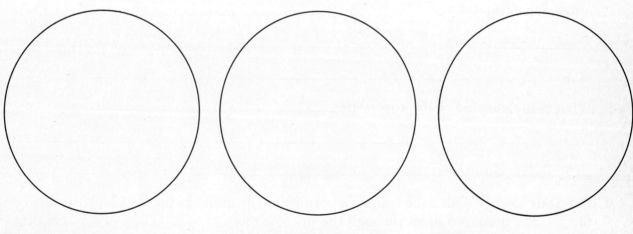

Newspaper Sample Under High Power Magnification _____

Magazine Sample Under High Power Magnification _____

Magazine Sample Under High Power Magnification _____

7. Move the slide to the left, then to the right.
8. Looking at the stage and objectives from the side, turn the nosepiece until the high-power objective clicks into place.
9. Using only the fine adjustment knob, bring the letter into focus. Draw what you see.
10. Repeat steps 2 through 9 using magazine samples.

Observations
1. While looking through the microscope, in which direction does the object appear to

   move when you move the slide to the left? _____

   To the right? _____

2. What is the total magnification of your microscope under low and under high

   power? _____

3. What happens to the focus of the objective lens when you switch from low power to

   high power? _____

   _____

Analysis and Conclusions
1. What conclusion can you draw about the way objects appear when viewed through a

   microscope? _____

   _____

   _____

   _____

2. How would you center an object viewed through a microscope when the object is

   off-center to the left? _____

   _____

   _____

3. What is the purpose of the coverslip? _____

   _____

   _____

4. **On Your Own** With your teacher's permission, examine samples of hair, various fabrics, and prepared slides through the microscope.

Answer Key

Chapter Discovery: Discovering Cell Function

1. protects the cell from the environment; keeps unwanted materials from the cell
2. allows necessary materials to enter the cell
3. obtains and stores energy 4. processes food for use by the cell 5. stores food and water 6. provides power for the cell
7. controls and governs the activities of the cell 8. repairs cell parts 9. builds new cell parts 10. gets rid of wastes
11. reproduction 12. transports materials

Critical Thinking and Application

1. Growth and repair of cell parts; controlling materials that enter and leave the cell; obtaining and storing energy; processing food; storing food and water; producing power; controlling cell activities; reproduction; getting rid of wastes; transporting materials.
2. Accept all reasonable answers; the preferred answer is that cells have specific structures that carry out specific functions.
3. For now, allow students to speculate freely. Possible answers include: cells in glands produce certain enzymes or hormones; cells designed to produce certain proteins; white blood cells produce antibodies.

Activity: Can You Identify These Cell Structures?

1. mitochondrion 2. cell wall
3. chromosome 4. nucleus 5. chloroplast
6. endoplasmic reticulum 7. cell membrane 8. ribosome 9. vacuole
10. lysosome

Activity: Shapes of Cells

Most school libraries have reference books and A-V materials, which are written on the appropriate reading level for students, concerning cells. If not, have your students use the public library. Impress upon your students that the shape of a cell is related to its function. For example, muscle cells are usually long for contracting, red blood cells are concave for carrying oxygen, and xylem cells are tubelike for transporting water and minerals.

Activity: Miracles of Modern Medicine

Check time lines to be sure that the letters are written in the correct places. From left to right the letters should appear in this order: J, H, B, E, G, A, D, C, I, F.

Activity: Biologists Who Studied Cells

Mathias Schleiden (1804–1881), a German botanist, and **Theodor Schwann** (1810–1882), a German physiologist, were co-founders of the cell theory. Schwann's further investigations proved that the cell is the basis of animal and plant tissue. **Robert Hooke** (1635–1703) was an English scientist who observed the cells of a thin piece of cork by using a microscope. He coined the term cells.

Discovery Activity: Division of Labor

This activity introduces students to the concept of biological coordination by means of an analogy to the cooperation within a community of human beings. You may wish to point out later that such groups of like organisms, or populations, are often referred to as a sixth level of biological organization—beyond the five levels discussed.

Discovery Activity: Shape and Function of Cells

Using a magnifying glass, students can observe small plant parts, such as root hairs. However, students must use a microscope to observe the cell structure of plant parts. To prepare slides of the plant parts collected, specimens must be sliced thin enough to allow light to pass through them. Be sure your students observe leaf, root, and stem cells. Depending on the type of slide preparation, cross section or longitudinal section, the cells will vary in shape. Have reference books available that show diagrams of leaf, stem, and root sections so that students can identify and learn the functions of the cells they observe.

Problem-Solving Activity: Cell Mathematics

1. 50,000,000,000,000 **2.** 5,000,000
3. 500,000 **4.** 0.008 **5.** 0.000001 **6.** 55
7. 500 **8.** 49.86 **9.** 0.00625 6.25
10. 0.008 0.0005 0.00009 **11.** 2 4 32
1024 1,048,576 **12.** 42 doublings **13.** 29
days

Many students have difficulty writing and understanding very large and very small numbers. Problems 1 to 5 provide practice in this skill. Problems 8 and 9 introduce the term "micron." Students might also be interested in a unit that is even smaller than a micron, the nanometer. A nanometer is one-thousandth of a micron (1 nm = .001 micron). Problem 10 involves the concept of doubling. This can be a fascinating concept for students to explore. It illustrates how environmental problems such as overpopulation can "creep up on us" overnight. The doubling of a population, or any other quantity, seems insignificant for several generations. All at once the population grows by "leaps and bounds." The answer to problem 11 will undoubtedly raise some feelings of disbelief. But, if students work out the problem step by step, they will discover the following: 8th doubling—approximately 2.54 cm thick; 12th doubling—slightly over 0.3 m thick; 20th doubling—about 104 m thick; 35th doubling—4828 km thick; and 42nd doubling—approximately 386,232 km thick.

Laboratory Investigation: Using the Microscope

Observations 1. When you move the slide to the left, objects appear to be moving to the right, and vice versa. **2.** Answers will vary depending on the microscopes used in your classroom. Explain to students that they should multiply the power of the eyepiece by the power of the lens to obtain the total magnification. **3.** When you go from low to high power, the object will no longer be in focus. You must refocus the microscope.

Analysis and Conclusions 1. Students should be able to conclude that objects appear larger, reversed, and upside down. **2.** Move it slightly to the right. **3.** A coverslip prevents the objects being viewed from moving and provides protection for both the object and the microscope lens. **4.** Have students make drawings of the samples they examine. Then discuss the similarities and differences among the observations students have made.

Contents

Chapter Discovery

Chapter Activities

Laboratory Investigation Worksheet

   (**Note:** *This investigation is found on page D76 of the student textbook.*)

*Appropriate for cooperative learning

Chapter Discovery

Discovering Cell Division

Materials
large sheet of posterboard
marking pen
scissors
colored yarn: red, white, black, yellow, green, blue, pink, brown
masking tape

Procedure

1. On the sheet of posterboard, draw three circles about 15 cm in diameter, as shown in Figure 1. Label these circles A, B, and C.

2. Draw a line through the center of circle C; then draw a line perpendicular to that line. (See Figure 1.)

3. Beneath the three circles, draw an oblong shape as shown. Label this shape D. Draw a line from one end of the oblong to the other. Then draw two circles side by side. Label these circles E.

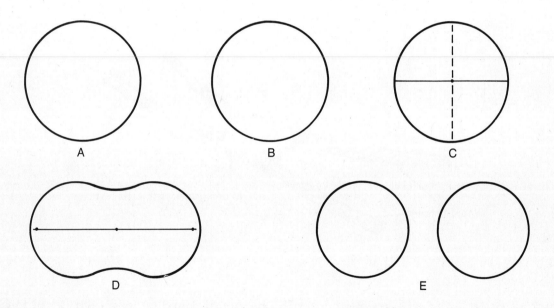

Figure 1

4. Cut two strands from each color yarn. Each strand should be about 10 cm long. When you are finished, you should have 16 strands, two of each color.

5. Scramble the 16 strands of yarn so that the colors are all mixed. Place the yarn inside circle A as shown in Figure 2.

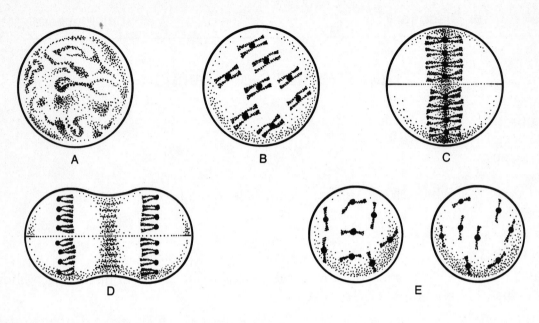

A   B   C

D   E

Figure 2

6. Separate the strands of yarn according to color.

7. Take a small piece of masking tape and wrap it around the middle of each strand pair as shown in Figure 3. Place each colored pair in circle B as shown in Figure 2.

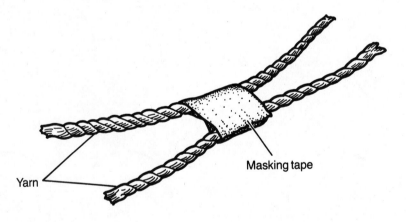

Masking tape

Yarn

Figure 3

8. Line up the pairs of yarn strands along the perpendicular line as shown in circle C of Figure 2.

9. Using the scissors, carefully cut each yarn strand pair in half through the middle of the masking tape. (See Figure 4.) As you do so, move each half of the pair to opposite ends of the oblong shape. (See Figure 2.)

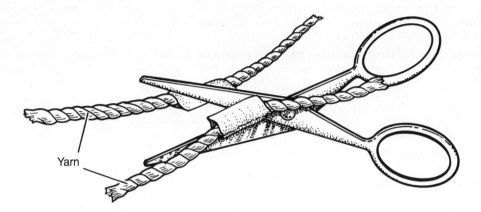

Yarn

Figure 4

10. Move one set of yarn strands from each end of the oblong shape into each of the two circles labeled E. (See Figure 2.)

Critical Thinking and Application

1. Each of the letters A through E represents a stage in the life of a cell. What do you see happening to the cell from the first stage through the fifth stage?

_____

_____

_____

_____

2. The strands of colored yarn represent genetic material in the cell. How does this genetic material appear in the cell in stage A? _____

_____

_____

3. What happens to the strands of genetic material in stages B and C?

_____

_____

_____

Cells: Building Blocks of Life D ■ 49

4. What has happened to the strands of genetic material by the time they reach

stage D? _____

_____

_____

_____

5. Describe the genetic material inside each of the circles in stage E. What do these circles represent? How does the genetic material in the two circles compare?

_____

_____

_____

_____

Activity

Associating Cells With Organisms and Events

Examine each of the cell drawings below. Determine whether the cell is associated with drawing A or drawing B. Circle the letter of your choice. On the lines provided, tell why you chose A or B.

1. Cell after mitosis

Which is the parent cell?

A.   B.

_____

_____

_____

2. Nerve cell

From which organism does this cell come?

A.   B.

_____

_____

_____

3. Chloroplast

Which part of the tree contains cells with this structure?

A.    B.

_____

_____

_____

4. Cell

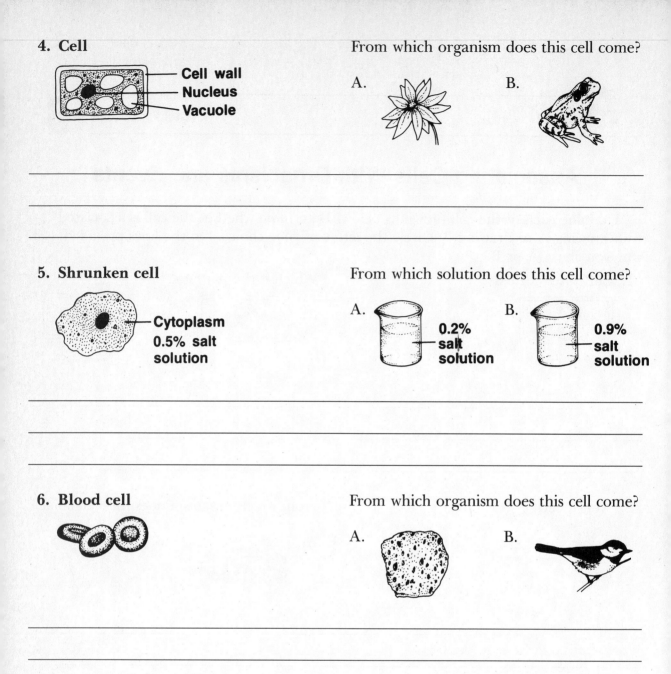

Cell wall
Nucleus
Vacuole

From which organism does this cell come?

A.

B.

_____

_____

_____

5. Shrunken cell

Cytoplasm
0.5% salt
solution

From which solution does this cell come?

A.
0.2% salt solution

B.
0.9% salt solution

_____

_____

_____

6. Blood cell

From which organism does this cell come?

A.

B.

_____

_____

_____

Find the Cell Phrase

On the lines below, write the word or words that best fit the description on the left. When you are finished, the boxed-in letters will spell out one of the topics discussed in the chapter. Fill in the word or phrase in the space provided.

1. Stage of mitosis in which
   chromosomes split apart   _□_____

2. Process by which a cell releases
   energy from food   ___□_____

3. Cell division and chromosome
   duplication   _____□_

4. A living thing   _____□__

5. Period between one mitosis and the
   next   ___□_____

6. Passage of water through cell
   membranes   __□____

7. Final stage of mitosis   __□____

8. Stores information that helps cells
   make the substances they need   _____□ _____

9. The second level of organization in
   a multicellular organism   ____□_

10. Cell structure in which sugars are
    broken down   _____□__

11. Organelle that contains enzymes
    for digestion   ____□___

12. Thin, flexible envelope surrounding
    a cell   ___ □_____

13. Nonliving material that makes up
    the cell walls of plant cells   ____□____

14. Process by which living things give
    rise to the same type of living thing ___□_____

Cells: Building Blocks of Life   D ■ 53

15. Used to build and repair body cells   _ _ _ _ ☐ _ _

16. Green pigment in plant cells   _ _ _ _ _ _ ☐ _ _ _ _ _ _

17. Control center of the cell   _ _ _ ☐ _ _ _

18. Groups of different tissues working
    together   _ _ _ ☐ _

19. All the living materials in a cell   _ _ _ ☐ _ _ _ _ _ _

20. Storage tank of the cell   _ _ ☐ _ _ _ _ _

An organelle   _ _ _ _ _ _ _ _ _ _ _   _ _ _ _ _ _ _ _ _ _ _

Activity

Observing the Effects of Diffusion in a Living Cell

By performing the following investigation, you can see for yourself how water moves into and out of plant cells.

Procedure

1. Remove a leaf from the tip of an *Elodea* plant, a common plant used in an aquarium. Place the leaf on a clean glass slide.

2. Use a medicine dropper to add a drop of fresh water to the slide. Carefully place a coverslip over the leaf.

3. Examine the leaf under the low-power objective of your microscope. On a separate sheet of paper, draw one cell and label the parts you can observe.

4. With the medicine dropper, add one drop of a salt solution to the edge of the coverslip.

5. Gently touch a paper towel to the opposite side of the coverslip. The paper towel will draw the salt solution over the leaf.

6. Examine the *Elodea* leaf again under the low-power objective of the microscope. Draw one cell and label the parts.

Critical Thinking and Application

1. How did the cells you observed in step 3 compare with those in step 6?

_____

_____

_____

_____

_____

2. Use the term diffusion to explain the differences you observed in the cells after the salt solution was added.

_____

_____

_____

_____

_____

3. What is the name of the special type of diffusion that involves water?

_____

_____

_____

_____

4. Before the invention of refrigeration, some types of food were preserved by adding large amounts of salt to them. What hypothesis can you offer to explain how the

addition of salt can preserve food? _____

_____

_____

_____

_____

_____

_____ *Laboratory Investigation* _____

Observing Mitosis

Problem

How do the phases of mitosis appear under a microscope?

Materials *(per group)*

prepared slides of mitosis in animal and plant cells
microscope

Procedure ▲

1. Begin your investigation by observing prepared slides of mitosis in animal cells.
2. Examine the slides in the order that corresponds to the phases of mitosis.
3. Draw and label what you observe on each slide.
4. Now observe prepared slides of mitosis in plant cells in the order that corresponds to the phases of mitosis.
5. Again draw and label what you observe.

Observations

Phase of Mitosis	Animal Cells	Plant Cells
_____		
_____		
_____		
_____		

Analysis and Conclusions

1. Based only on your observations, compare mitosis in animal and plant cells.

   _____

   _____

   _____

   _____

   _____

   _____

2. Were there events in mitosis you read about in this chapter that you could not

   observe under the microscope? _____ If so, which events were they? _____

   _____

   _____

   _____

   _____

3. Mitosis is only one part of the cell division process. Did any of your slides show other

   phases of cell division? _____ If so, what were they? _____

   _____

   _____

   _____

4. **On Your Own** Using materials of your choice, construct a three-dimensional model of mitosis in a plant cell or in an animal cell.

Answer Key

Chapter Discovery: Discovering Cell Division

Critical Thinking and Application **1.** The cell divides into two cells. **2.** It is scrambled together in no particular order. **3.** Pairs are separated. The strands of each pair are joined at the center; then the pairs line up along the center of the cell. **4.** Each pair has split in half at the center. One half of the pair has gone to one end of the cell; the other half of the pair has gone to the other end of the cell. **5.** The circles represent two new cells; the genetic material in each new cell is like that in the original cell; the genetic material in each of the two cells is alike.

Problem-Solving Activity: Associating Cells With Organisms and Events

1. A. Mitosis produces daughter cells with the same number and kind of chromosomes as the parent cell. **2.** B. Nerve cells are not found in plants. **3.** B. Chloroplasts are generally found in the leaves or green parts of plants. **4.** A. Cell walls are found only in plant cells. **5.** B. A higher concentration of salt outside the cell (hypertonic solution) causes water to move out of the cell. **6.** B. Although the sponge is an animal, it does not contain blood.

Activity: Find the Cell Phrase

1. anaphase **2.** respiration **3.** mitosis **4.** organism **5.** interphase **6.** osmosis **7.** telophase **8.** nucleic acid **9.** tissue **10.** mitochondrion **11.** lysosome **12.** cell membrane **13.** cellulose **14.** reproduction **15.** protein **16.** chlorophyll **17.** nucleus **18.** organ **19.** protoplasm **20.** vacuole **Phrase** endoplasmic reticulum

Discovery Activity: Observing the Effects of Diffusion in a Living Cell

This activity will enable students to observe the results of diffusion of salt water through the cell membrane of a fresh-water plant, *Elodea*. You may even want to introduce students to a new word—plasmolysis—the shrinkage of cell contents because of a loss of water through the cell membrane. Students will be able to observe plasmolysis of cells of the *Elodea* leaf placed in salt solution. The cell contents shrink, but because the cell wall is rigid, the plant cell retains its original shape. If the cell is left in salt water, plasmolysis continues and the cell dies from lack of water. Students should be able to explain their observations using the term diffusion and the term osmosis—the diffusion of liquids through a cell membrane. **Critical Thinking and Application** **1.** The cell membrane in the cells in step 6 has pulled away from the cell wall. The contents of the cell look more dense since they have been squeezed into a smaller space by the contraction of the cell membrane. **2.** The higher concentration of salt in the solution added to the cell in step 6 caused water molecules to diffuse out of the cell. There are fewer water molecules in the salt solution than in the cell. The water molecules moved from an area of higher concentration to an area of lower concentration. **3.** Osmosis is the special type of diffusion that involves water. **4.** When salt is added to food, water diffuses from the cells. Thus, the food is preserved because the bacteria and fungi that normally cause food to spoil cannot live in cells that have been dried. Salt can also cause water to diffuse out of bacteria cells that land on food. The bacteria die.

Laboratory Investigation: Observing Mitosis

Observations Check students' drawings for scientific accuracy. From top to bottom, phases are: Prophase, Metaphase, Anaphase, Telophase. **Analysis and Conclusions** **1.–3.** Answers will vary, depending on the content of the prepared slides provided to students. Mitosis is basically the same in plant and animal cells. One difference is that centrioles appear during prophase in animal cells but not in plant cells. The final stage of cell division, cytokinesis, is quite different in plant and animal cells. In animal cells, a depression called a cleavage furrow forms, and eventually, the cytoplasm is cut in two.

The cells of most land plants have fairly rigid walls that do not lend themselves to forming cleavage furrows. Instead, cytokinesis is accomplished through cell-plate formation. The cell plate then becomes part of the cell wall, a structure found in plant cells but not in animal cells.   **4.** Check students' models for scientific accuracy.

Contents

*Appropriate for cooperative learning

Discovering the Products of Photosynthesis

Materials
leaf from a geranium plant
double boiler
solution of iodine mixed with water
medicine dropper
alcohol
shallow dish
forceps or kitchen spoon
hot plate

Procedure

This activity must be performed under a teacher's or guardian's supervision.

1. Place the geranium leaf in the top of the double boiler. Cover the leaf with alcohol. **CAUTION:** *Alcohol burns easily and quickly.* Do not use a heat source with an open flame. NEVER put a container of alcohol directly over a flame.

2. Fill the bottom half of the double boiler with hot water and bring it to a boil.

3. Set the top of the double boiler, with the leaf and alcohol, over the boiling water in the bottom half.

4. Heat the leaf until it loses its color. Then turn off the heat source.

5. Use the spoon to remove the leaf and put it on the small dish.

Observation: What color is the alcohol in which the leaf was heated?

_____

6. Wash the leaf gently under tap water.

Observation: What color is the leaf?

_____

7. Put the leaf back on the dish and use the medicine dropper to cover it with iodine solution. Let the iodine stay on the leaf for several minutes.

8. Pour off the excess iodine and gently wash the leaf again in tap water. Wash off the plate and place the leaf back on the plate.

Observation: What color was the leaf once you covered it with iodine solution?

_____

Critical Thinking and Application

1. What caused the alcohol to turn green? Do you know what the green substance is
   called? _____

   _____

   _____

   _____

2. What do you think caused the leaf to turn blue-black when it was covered with the

   iodine solution? _____

   _____

   _____

   _____

3. Do you think there is a relationship between the green color of the leaf and the
   substance that turned blue-black in the presence of iodine? If so, what do you think

   that relationship is? _____

   _____

   _____

   _____

   _____

   _____

   _____

   _____

Activity

Analyzing Photosynthesis and Respiration

During photosynthesis, green plants use carbon dioxide and water to produce food in the form of glucose. During respiration, the glucose is broken down to be used as energy by the plant. As the glucose is broken down, oxygen is released by the plant. Carbon dioxide, oxygen, and water form a continuous cycle during these two processes.

Figure 1

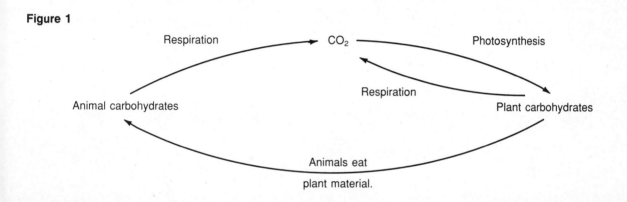

Study the diagram of the carbon cycle that is shown in Figure 1. Then answer the questions, based on the diagram and your knowledge of photosynthesis and respiration.

1. The concentration of CO_2 in the atmosphere remains at a stable 0.004 percent.

Which two processes keep this concentration stable? _____

_____

2. Plants depend upon the activities of animals for a continuing supply of which

substance? _____

_____

3. Which process removes CO_2 from the atmosphere? _____

4. Which process adds CO_2 to the atmosphere? _____

5. Into which organic compound does photosynthesis convert the carbon of CO_2?

_____

6. After plants are eaten by animals, what process changes the carbon in these organic

compounds back to CO_2? _____

Activity

Chlorophyll and Light

A pigment is a substance that absorbs and reflects different colors of light. (You might already know that "white" light contains all of the colors of the spectrum.) For example, the color of a leaf is due to the green pigment chlorophyll. When white light shines on chlorophyll, the chlorophyll absorbs most of the red, orange, blue, and violet and reflects most of the green and yellow. That is why you see a yellow-green color in a leaf. You can think of a pigment as a kind of sponge that soaks up all of the colors of the spectrum except the ones you see.

Use the bar graph below, which shows the percentage of light energy reflected by chlorophyll, to answer the following questions.

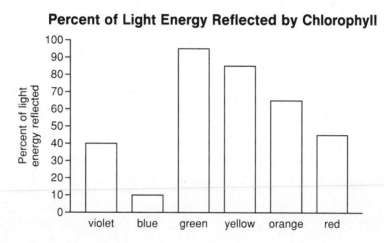

Percent of Light Energy Reflected by Chlorophyll

1. Which color in this spectrum is most visible? _____

_____

2. What is the approximate percentage of light reflected for this color? _____

3. What percentage of light energy absorbed does this represent? _____

4. If everything above 50 percent of light energy reflected is visible to the human eye, is red light part of the mixture of colors seen in light that is reflected by chlorophyll?

_____

_____

_____

Activity

Carbon Dioxide

Carbon dioxide is a common gas. Its formula, CO_2, tells you that it is made of the elements carbon and oxygen. Your body produces CO_2 when it "burns" the food you eat. When you exhale, your body gets rid of this CO_2.

Materials
50 mL beaker
30 mL limewater
straw
bottle
40 mL limewater
wire
candle
watch crystal
2 spoonfuls baking soda
30 mL vinegar
match

Procedure
A. To test for the CO_2 in your breath, pour 30 ml of limewater into a 50 mL beaker. Use a straw to blow bubbles through the limewater for about 30 seconds. What happens to

the limewater? _____

_____

When CO_2 combines with the limewater, $CaCO_3$ (calcium carbonate) is formed. $CaCO_3$ is a white substance that does not dissolve well in water. Scrub the beaker with a brush and soap and return it upside down to its proper place. Put the straw in the trash.

B. Get a bottle and pour 40 ml of limewater into it. Wrap the end of a wire around a candle. Also get a watch crystal. At your place, light the candle and lower it with the wire into the bottle as shown. Let it burn for about 10 seconds. Then put the watch crystal over the mouth of the bottle until the candle goes out. Slide the watch crystal back and remove the candle. Slide the watch crystal back over the mouth of the bottle

and shake it well. What happens to the limewater? _____

_____

This shows that the burning candle produced _____

_____.

Where did the oxygen come from to form the CO_2? (*Hint:* From where do you get

your oxygen?) _____

_____

Where did the carbon come from to form the CO_2? _____

_____

C. Empty the bottle into the sink. Put 2 spoonfuls of baking soda into the bottle. Pour 30 ml of vinegar into the 50 ml beaker. Pour the 30 ml of vinegar from the beaker into the bottle with the baking soda. Cover the bottle immediately with the watch crystal.

What happened when the vinegar hit the baking soda? _____

_____

_____

D. Light the candle and lower it into the bottle of gas. **Note:** *Don't get the candle wet!*

What happens? _____

_____

_____

Remove the candle from the bottle and quickly pour the gas (not the liquid) from the bottle into the 50 ml beaker. Quickly light a match and lower it into the beaker. What

happens? _____

_____

_____

Scrub and dry the watch crystal, beaker, and bottle.

_____ *Laboratory Investigation* _____

Comparing Photosynthesis and Respiration

Problem

What is the relationship between photosynthesis and respiration?

Materials *(per group)*

100-mL graduated cylinder   2 *Elodea*
bromthymol blue solution   2 #5 rubber stoppers
2 125-mL flasks   light source
straw

Procedure ⚠ ⚕

1. Using the graduated cylinder, pour 100 mL of bromthymol blue solution into each flask. **CAUTION:** *Bromthymol blue is a dye and can stain your hands and clothing.*

2. Insert one end of the straw into the bromthymol blue solution in one of the flasks. Gently blow through the straw. Keep blowing gently until there is a change in the color of the solution. (Bromthymol blue turns yellow in the presence of carbon dioxide.) Repeat this procedure with the other flask.

3. Place a sprig of *Elodea* into each flask. *Elodea* is an organism that will perform photosynthesis when placed in sunlight. Put a stopper in each of the flasks.

4. Place one flask in the dark for 24 hours and the other flask on a sunny windowsill for the same amount of time.

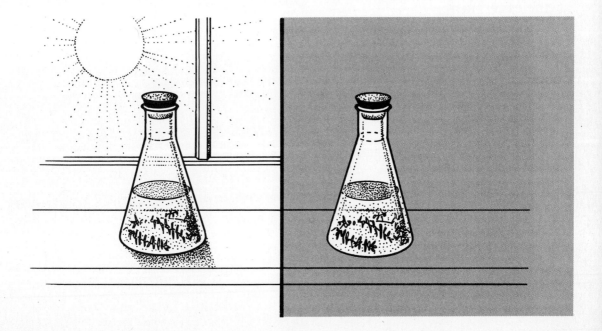

Observations

1. What was the color of the bromthymol blue solution before and after you exhaled into it? _____

2. What was the color of the bromthymol blue solution in the flask placed in the dark for 24 hours? In the flask on the windowsill? _____

Analysis and Conclusions

1. How was the carbon dioxide you exhaled into the bromthymol blue solution produced in you body? _____

_____

_____

2. Why was *Elodea* placed in both flasks? _____

_____

_____

3. How can you explain your observations for each flask? _____

_____

_____

4. How are photosynthesis and respiration related? _____

_____

_____

5. **On Your Own**   Design and experiment to see if temperature changes have any effect on the results of this experiment. With your teacher's permission, carry out the experiment.

Answer Key

Chapter Discovery: Discovering the Products of Photosynthesis

5. The alcohol is green.   **6.** The leaf is a grayish color—obviously without pigment. **8.** The leaf is a blue-black color. **Critical Thinking and Application   1.** Students should realize that the green color of the leaf caused the alcohol to turn green. This substance is chlorophyll.   **2.** Some students may know that starch turns blue-black in the presence of iodine.   **3.** At this time, allow students to speculate. They should have some idea of the relationship that exists between chlorophyll, photosynthesis, and starch. The correct answer to this question is that the green color in the leaf is chlorophyll, the substance that uses light energy used in the leaf's food-making process. A leaf stores the food it makes in the form of starches. Once the green color of the leaf is removed by the alcohol, the addition of the iodine solution causes the starch stored in the leaf to become visible.

Activity: Analyzing Photosynthesis and Respiration

1. Respiration and photosynthesis keep this concentration stable.   **2.** Plants depend upon animals for a continuing supply of CO_2. **3.** Photosynthesis removes CO_2 from the atmosphere.   **4.** Respiration adds CO_2 to the atmosphere.   **5.** In photosynthesis, the carbon in CO_2 is used to make glucose. **6.** Respiration changes the carbon in these compounds back to CO_2.

Activity: Chlorophyll and Light

1. green   **2.** 94 percent   **3.** 6 percent **4.** No.

Discovery Activity: Carbon Dioxide

Procedure A. It turns milky. **B.** It turns milky. Carbon dioxide. The air. Organic materials. In the case of the burning candle, petroleum products from which the candle was made and that were present in the burning wick. **C.** Bubbles were given off. **D.** The match went out.

Laboratory Investigation: Comparing Photosynthesis and Respiration

Observations   1. The solution was blue before exhaling and yellow after exhaling. **2.** The solution in the dark remained yellow, and the solution on the windowsill turned blue.   **Analysis and Conclusions   1.** The carbon dioxide was produced at the cellular level by the process of respiration.   **2.** *Elodea* was placed in both flasks to test for the presence of oxygen via photosynthesis. **3.** The solution kept in the dark remained yellow because *Elodea* does not perform photosynthesis in the dark. The solution exposed to light turned blue in color because the process of photosynthesis in *Elodea* released oxygen into the solution   **4.** The waste product oxygen, released by photosynthesis, allows oxygen-breathing organisms to survive. The waste product carbon dioxide, released through the process of respiration in oxygen-breathing organisms, allows photosynthesis to occur in plants. **5.** Experiments will vary, depending on the student. Each experiment, however, should be designed to test the effect of temperature changes on photosynthesis and respiration.

Science Reading Skills

TO THE TEACHER

One of the primary goals of the *Prentice Hall Science* program is to help students acquire skills that will improve their level of achievement in science. Increasing awareness of the thinking processes associated with communicating ideas and reading content materials for maximum understanding are two skills students need in order to handle a more demanding science curriculum. Teaching reading skills to junior high school students at successive grade levels will help ensure the mastery of science objectives. A review of teaching patterns in secondary science courses shows a new emphasis on developing concept skills rather than on accumulating factual information. The material presented in this section of the Activity Book serves as a vehicle for the simultaneous teaching of science reading skills and science content.

The activities in this section are designed to help students develop specific science reading skills. The skills are organized into three general areas: comprehension skills, study skills, and vocabulary skills. The Science Gazette at the end of the textbook provides the content material for learning and practicing these reading skills. Each Science Gazette article has at least one corresponding science reading skill exercise.

Contents

Name _____ Class _____ Date _____

Fernando Nottebohm: From Bird Songs to Reborn Brains
Science Reading Skill: Defining Technical Terms

Perhaps you have had difficulty with some of the terms that you have come across in reading science material. One skill that can help you find the meaning of scientific terms is using context clues. In this technique, you find a phrase or sentence that describes, gives an example of, or states the meaning of the word. From this phrase or sentence comes a definition of the word.

Part A

Listed below are several scientific terms used in this article. Use the skill of finding word meanings from context clues to write a definition of each term. Write your definition on the blank lines provided next to each word. If the word is new to you and you cannot determine the meaning from its context, refer to the glossary in the back of the science textbook or to a dictionary. Be sure to find the scientific meaning of the term as it is used in the article.

1. neurogenesis _____

_____

2. neurons _____

_____

3. regenerate _____

_____

4. cells _____

_____

5. membrane _____

_____

6. syrinx _____

_____

7. vibrates _____

_____

Part B

For each of the following words, write a sentence of your own. Be sure to use the meaning of the word as it appears in the article.

1. neurogenesis _____

_____

2. neurons _____

_____

3. regenerate _____

_____

4. cells _____

_____

5. membrane _____

_____

6. syrinx _____

_____

7. vibrates _____

_____

Science Reading Skill: Sequence of Events

The following information is a list of events that are described in the selection. After reading all of the events, arrange them in the order in which they occur. Write the letter of the event on the blank line next to the number that represents its sequence. In other words, put the letter of the event that occurred first on the blank next to the number one, and so on.

1. _____ a. Fernando Nottebohm and Arthur Arnold discover that neurogenesis occurs in adult birds.

2. _____ b. Fernando becomes interested in birds' songs as he wanders on his family's ranch.

3. _____ c. Fernando studies agriculture.

4. _____ d. Fernando and his partner find out that the size of the song-controlled areas in the brains of male canaries changes dramatically with the seasons.

5. _____ e. Fernando studies how a membrane in a bird's voice box vibrates when the bird sings.

Human Growth Hormone—Use or Abuse?
Science Reading Skill: Fact and Opinion

In reading science material, it is important to be able to tell the difference between a statement of fact and an opinion. The truth of a statement that appears to be factual can be determined by checking with authorities in the field or checking appropriate reference materials. A statement of opinion conveys what a person thinks or believes about something. A theory is an explanation that has not yet been proven as factual.

After reading this article, try your skill in identifying statements of fact or opinion. Write "O" for opinion or "F" for fact on the line next to each statement.

_____ 1. Marco wants to be tall more than anything else.

_____ 2. HGH is normally produced in the body.

_____ 3. Synthetic HGH is almost identical to natural HGH.

_____ 4. Taking a growth hormone produced by bacteria is unsafe.

_____ 5. The cost of HGH injections is too high.

_____ 6. Some children have a deficiency of HGH.

_____ 7. Drug companies don't care about anything but making money.

_____ 8. Variations in height are natural.

_____ 9. Marco does not have a deficiency of HGH.

_____ 10. All the side effects of taking HGH are not yet known.

_____ 11. Being tall is better than being short.

_____ 12. Marco will be happier as an adult because he took HGH.

Science Reading Skill: Critical Reading

Critical reading is careful, thoughtful reading. Some of the important skills involved in critical reading are examining and evaluating the facts, making inferences, and understanding word meanings from context.

A critical reader examines the facts to see how they relate to the topic of the article. Understanding this relationship will help make clear the author's main point. In evaluating the facts, the reader compares these facts to what is already known. An inference is a reasonable conclusion a reader comes to as a result of a suggestion the author makes. An inference is an idea not directly stated in the reading material. A critical reader tries to get the meaning of an unfamiliar word from the way it is used in a sentence or a paragraph and from the other familiar words surrounding it. Finally, a critical reader thinks along with the author while reading.

You should practice using critical reading skills as often as you can. The questions below deal with the article you have just read. For each question, circle the letter of the correct choice or write your answer on the lines provided.

1. Why do you think HGH was not so widely used in the past?
   a. It was too dangerous.
   b. It was too expensive.
   c. Doctors recommended against it.
   d. Doctors were not asked about it.

2. What is the difference between Marco and a child who has a deficiency of HGH?

   _____

   _____

   _____

3. Which of the following are reasons a doctor might give for not giving HGH to a child who does not have a deficiency of HGH? Circle two.
   a. Tall people don't live as long as short people.
   b. HGH is very expensive.
   c. It can only be taken for a few months.
   d. There are side effects.

4. For whom was HGH developed? _____

   _____

   _____

5. Why might drug companies want HGH to be taken by other children as well?

   _____

   _____

   _____

   _____

6. A hormone is something that
   a. the body produces.
   b. is contained in most drugs.
   c. can be found only in bacteria.
   d. some humans produce.

7. One reason that some people want to take HGH is that they think
   a. taller is stronger.
   b. shorter is healthier.
   c. people should be as intelligent as possible.
   d. appearances are important.

8. What does HGH stand for?
   a. high gestation halogen
   b. happy-go-high
   c. human growth hormone
   d. here goes height

9. Variety in human traits is
   a. normal.
   b. rare.
   c. a problem.
   d. a new phenomenon.

10. Which of the following would make another good title for this article?
   a. How Synthetic HGH Is Produced
   b. How Tall and at What Cost?
   c. A New Way to Achieve a High Position
   d. Can HGH Be Cost Effective?

The Green People of Solaron
Science Reading Skill: Understanding a Word's Meaning From Its Context

As you read in science, you may come across words that you have never encountered before. For example, the word "anthocyanin" may be quite unfamiliar to you. If you have a dictionary at hand, you can look up the meaning of the word. But if a dictionary is not available, you can still determine the word's meaning by using a different technique. You can get the meaning from the way the word is used—that is, from its context. By reading this article, you will discover that anthocyanin is a chemical that causes the leaves of plants and trees to turn red, blue, and purple.

When you are trying to figure out the meaning of an unknown word, look at the way the word is used in the reading material. Sometimes the word is specifically defined in another phrase or sentence. Or sometimes the word is described well enough in the reading material that you can understand its meaning. In this exercise, you will practice the technique of understanding a word's meaning from its context.

The words listed below come from this article. Under each word are some possible meanings. Select the correct meaning of the word from the context of the sentence(s) in which the word appears. Circle the letter of the correct answer.

1. horizon
   a. upper layer of soil
   b. point where the sky seems to meet the land
   c. peak of a mountain
   d. lines on a map running east-west

2. gape
   a. to stare with the mouth open
   b. to enclose in a bubble
   c. to color green
   d. to make oxygen

3. photosynthesis
   a. process of developing film
   b. process of making synthetic fibers
   c. process of growing food on farms
   d. process by which sunlight is used to make glucose from carbon dioxide and water

4. glucose
   a. sugar used as food
   b. fat made by plants
   c. green pigment
   d. special house with clear plastic roof

5. genetic engineering
   a. the process of obtaining stored glucose from the liver
   b. the process of making food
   c. the technique of copying genes
   d. the process of building a space colony

6. viruses
   a. skin cells
   b. new form of life found on Solaron
   c. scientists who study photosynthesis
   d. cause of the common cold

7. chlorophyll
   a. red pigment in leaves
   b. special green chemical involved in photosynthesis
   c. chlorine-containing substance
   d. clear plastic used for space colony bubble

8. solar
   a. pertaining to the sun
   b. referring to the moon
   c. having to do with space travel
   d. denoting green coloration

9. pigment
   a. extremely small person
   b. made-up story
   c. stored sugar
   d. coloring substance

10. volunteers
    a. inhabitants of Solaron
    b. plants that grow into trees
    c. people who offer to participate in an experiment
    d. skin cells that undergo photosynthesis

Adventures in Science

Defining Technical Terms   **Part A**
1. the birth of new neurons   **2.** nerve cells in the brain   **3.** grow back   **4.** small units that make up living things   **5.** thin, soft layer of tissue   **6.** song box   **7.** moves back and forth rapidly.   **Part B** Answers will vary.
Sequence of Events b, c, e, d, a

Issues in Science

Fact and Opinion   **1.** O   **2.** F   **3.** F
4. O   **5.** O   **6.** F   **7.** O   **8.** F

9. F   **10.** F   **11.** O   **12.** O   **Critical Reading**   **1.** b   **2.** Marco does not have a medical condition; the other child does.   **3.** b, d   **4.** Children who do not produce adequate amounts of HGH.   **5.** Few children have an HGH deficiency.   **6.** a   **7.** d   **8.** c   **9.** a   **10.** b

Futures in Science

Understanding a Word's Meaning From Its Context   **1.** b   **2.** a   **3.** d   **4.** a   **5.** c   **6.** d   **7.** d   **8.** a   **9.** d   **10.** c

Activity Bank

TO THE TEACHER

One of the most exciting and enjoyable ways for students to learn science is for them to experience it firsthand—to be active participants in the investigative process. Throughout the *Prentice Hall Science* program, ample opportunity has been provided for hands-on, discovery learning. With the inclusion of the Activity Bank in this Activity Book, students have additional opportunities to hypothesize, experiment, observe, analyze, conclude, and apply—all in a nonthreatening setting using a variety of easily obtainable materials.

These highly visual activities have been designed to meet a number of common classroom situations. They accommodate a wide range of student abilities and interests. They reinforce and extend a variety of science skills and encourage problem solving, critical thinking, and discovery learning. The required materials make the activities easy to use in the classroom or at home. The design and simplicity of the activities make them particularly appropriate for ESL students. And finally, the format lends itself to use in cooperative-learning settings. Indeed, many of the activities identify a cooperative-learning strategy.

Students will find the activities that follow exciting, interesting, entertaining, and relevant to the science concepts being learned and to their daily lives. They will find themselves detectives, observing and exploring a range of scientific phenomena. As they sort through information in search of answers, they will be reminded to keep an open mind, ask lots of questions, and most importantly, have fun learning science.

Contents

Activity

Activity _____

Hydra Doing?

Hydras are members of the phylum *Cnidaria*. Like all cnidarians, hydras have soft bodies made up of two layers of cells. They also have stinging tentacles arranged in circles around their mouth. Do hydras have the characteristics that all forms of life share? To answer this question, you will need a medicine dropper, a culture of hydras, a depression slide, a microscope, a toothpick, and some fish food.

Procedure 🔺 🐁

1. Using a medicine dropper, remove a drop of the hydra culture from the bottom of the culture jar.

2. Place the drop on a depression slide.

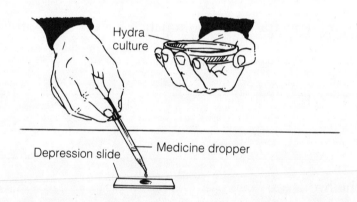

3. Put the slide on the stage of a microscope. Using the low-power objective, locate a hydra. Draw and label what you observe.

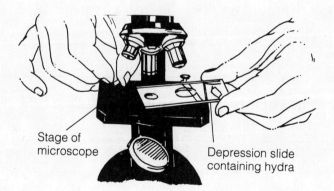

4. Carefully touch the hydra's tentacles with a toothpick. Observe what happens.

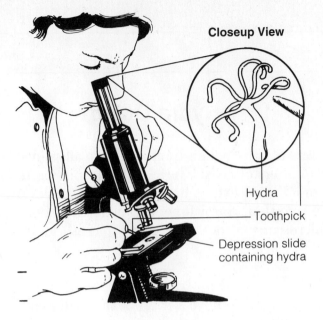

Closeup View

Hydra

Toothpick

Depression slide containing hydra

5. Place a tiny amount of fish food near the hydra and see what happens.

Observations

1. What happened to the hydra when you touched it with the toothpick?

_____

_____

2. How does the hydra move? _____

_____

3. Describe how the hydra eats. _____

_____

Analysis and Conclusions

1. What characteristics of living things does the hydra exhibit? _____

_____

_____

2. Does the hydra react as a whole organism or does just part of the hydra react?

Explain. _____

_____

_____

Going Further

Place a drop of vinegar (weak acid) near a hydra. Observe what happens.

— **A**ctivity ————————————————————————————————— CHAPTER

The Nature of Life **1**

What Is It?

The most common organic compounds found in living things are carbohydrates, proteins, and fats. Most foods contain combinations of these organic compounds. How can you identify the presence of carbohydrates (starches and sugars), proteins, and fats in foods? Try this activity to find out.

Materials

3 medicine droppers
iodine solution
slice of raw potato
hot-water bath
graduated cylinder
honey solution
2 test tubes
Benedict's solution
egg-white mixture
Biuret solution
vegetable oil
brown paper bag
paper towel
test-tube rack

Procedure

Testing for Starch ⚗ ☣

1. Using a medicine dropper, add a drop of iodine solution to a slice of raw potato.

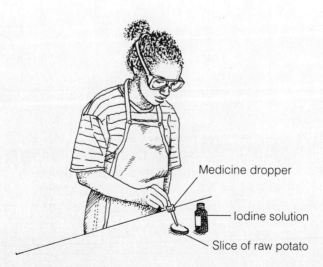

Medicine dropper

Iodine solution

Slice of raw potato

2. Record your observations in Data Table 1. The blue-black color indicates the presence of starch.

DATA TABLE 1

Substance	Color at Start	Color After Adding Iodine Solution	Starch Present?
Potato			

Testing for Sugar 🜃 🔥 🧰 👉 👁

1. Set up a hot-water bath as shown in the diagram. **CAUTION:** *Be careful when heating substances.*

2. While the water bath is heating, add 5 mL of honey solution to a test tube. Then add 10 mL of Benedict's solution to the test tube.

3. Gently swirl the contents of the test tube.

4. Place the test tube in the hot-water bath for about 5 minutes. After 5 minutes, observe the color of the mixture. If small amounts of sugar are present, the color of the mixture changes to green or yellow. If larger amounts of sugar are present, the color of the mixture changes to orange or red. Record your observations in Data Table 2.

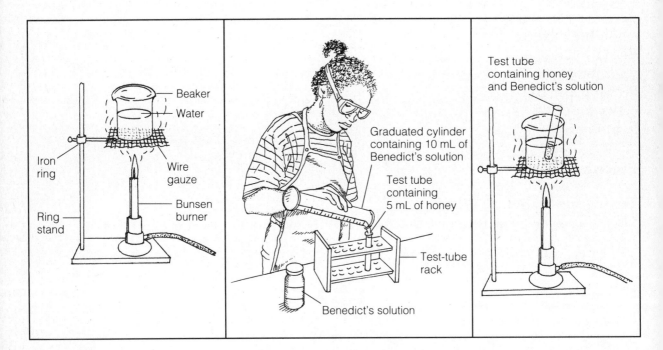

DATA TABLE 2

Substance	Color at Start	Color After Adding Benedict's Solution and Heating	Sugar Present?
Honey solution			

Testing for Protein 🧪 ⚗ 👁

1. Place 5 mL of an egg-white mixture in a test tube.

2. Using a clean medicine dropper, add 5 drops of Biuret solution. **CAUTION:** *Be careful when using Biuret solution. It may burn your skin and clothes.*

3. Gently swirl the contents of the test tube. If the color of the mixture changes to purple, protein is present. The darker the purple, the more protein there is. Record your observations in Data Table 3.

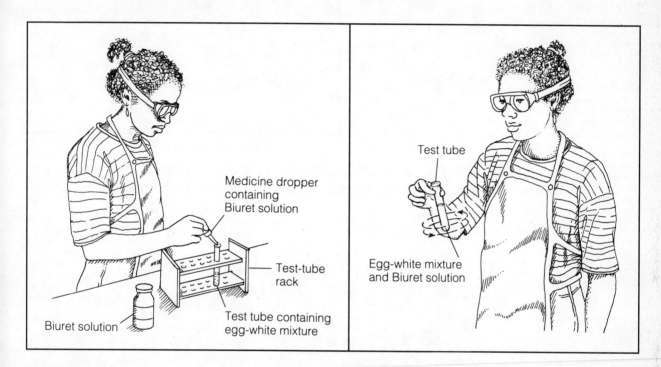

Medicine dropper containing Biuret solution

Test-tube rack

Biuret solution

Test tube containing egg-white mixture

Test tube

Egg-white mixture and Biuret solution

DATA TABLE 3

Substance	Color at Start	Color After Adding Biuret Solution	Protein Present?
Egg-white mixture			

Testing for Fat

1. Using a clean medicine dropper, add a few drops of vegetable oil to a piece of brown paper as shown in the diagram on the following page.

2. Rub off any excess oil with a paper towel. Set the paper aside for 10 minutes.

Medicine dropper

Vegetable oil

Piece of brown paper

Paper towel

3. After 10 minutes, hold the paper up to a bright light or a window. The translucent spot indicates the presence of fat. Record your observations in Data Table 4.

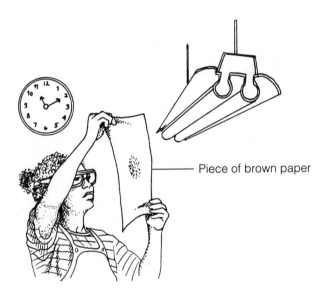

Piece of brown paper

DATA TABLE 4

Substance	Translucent Spot on Brown Paper	Fat Present?
Vegetable oil		

Analysis and Conclusions

1. Which substance contained starch? _____

   _____

   Sugar? _____

   _____

2. Which substance contained protein? _____

   _____

3. Which substance contained fat? _____

   _____

4. Your brown-paper lunch bag has a translucent spot on the bottom. What does this

   tell you about your lunch? _____

   _____

   _____

   _____

   _____

Going Further

   Repeat the activity on a variety of other foods. Which foods contain starches, sugars, proteins and/or fats?

ctivity

Now You See It—Now You Don't

Cells are the basic units of structure and function of all living things. Cells contain certain structures that carry out life processes. One such structure is the cell membrane. The cell membrane regulates what goes into and what comes out of a cell. In plants, the cell membrane is just inside the cell wall. For this reason, it is sometimes difficult to see the cell membrane. To help you see the cell membrane in plants a little better, why not try this activity.

Materials

2 medicine droppers
microscope slide
forceps
Elodea leaf
pencil
coverslip
microscope
salt solution
paper towel

Procedure 🔺

1. Using a medicine dropper, place one drop of tap water on a microscope slide.

2. With forceps, place an *Elodea* leaf in the drop of water. Cover with a coverslip.

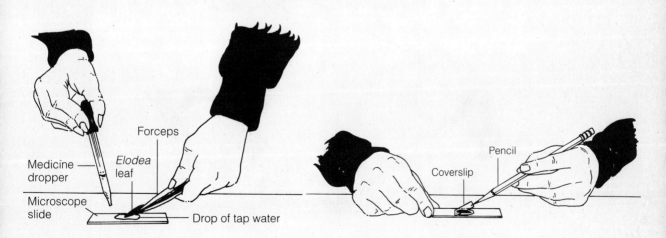

3. Observe the leaf under both low and high powers of the microscope. Note the location of the chloroplasts in relation to the cell wall. Draw a diagram of a plant cell, labeling the structures you observe.

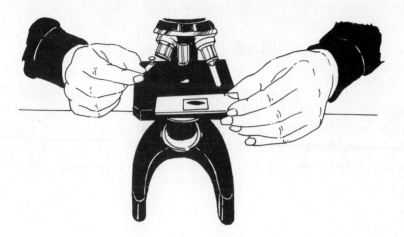

4. Using the other medicine dropper, add a drop of the salt solution along one edge of the coverslip. Place a piece of paper towel along the opposite edge of the coverslip as shown in the diagram. The tap water will soak into the paper towel, drawing the salt solution under the coverslip.

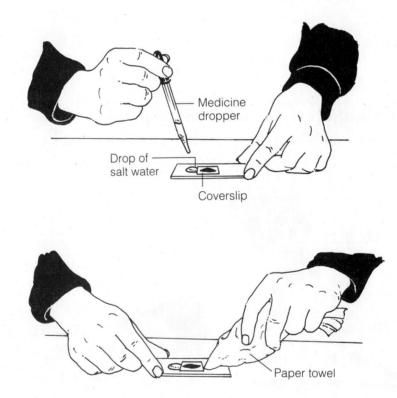

5. Observe the leaf under both low and high powers of the microscope. Again note the location of the chloroplasts in relation to the cell wall. Draw another diagram of the cell, labeling the cell wall, cell membrane, and chloroplasts.

6. Repeat step 4 using a drop of tap water instead of a drop of the salt solution. Observe the appearance of the cells.

Observations

1. Where were the chloroplasts located in the cell when tap water was added?

_____

_____

2. Where were the chloroplasts located when the salt solution was added?

_____

_____

_____

Analysis and Conclusions

1. In which direction did water move when the salt solution was added?

_____

_____

_____

_____

2. What happened to the cell when the salt solution was added?

_____

_____

_____

_____

3. What happened to the cell when tap water was added for the second time?

_____

_____

_____

_____

Going Further

Repeat this activity using a sugar (glucose) solution. Does a glucose solution have the same effect on a cell as a salt solution does?

Activity

Cell Processes

Coming and Going

One of the most important processes that occurs in a cell is diffusion. Diffusion regulates what enters and leaves the cell. To see how diffusion actually occurs, try this activity.

Materials

50 mL household ammonia
wide-mouthed jar
piece of cheesecloth (15 cm × 15 cm)
rubber band

spatula
gelatin "cell"
clock with second indicator

Procedure 🧪 ⚗ 👁

1. Select one member of the group to act as a Principal Investigator, a second member to act as a Timer, and a third member to act as a Recorder. The remaining members of the group will be the Observers. Be sure you understand your role in the activity before you continue.

2. Carefully pour 50 mL of ammonia in a wide-mouthed jar. **CAUTION:** *Keep the ammonia away from your skin and do not inhale its vapors. Keep the room well ventilated.*

3. Place the cheesecloth over the mouth of the jar. Hold it in place by putting a rubber band around the jar and the cheesecloth.

4. Using a spatula, place a gelatin "cell" on top of the cheesecloth as shown in the diagram. The gelatin "cell" contains a chemical called phenolphthalein. Phenolphthalein indicates the presence of bases such as ammonia. If a base is present, phenolphthalein will turn pink. **Note:** *Do not allow any ammonia to come into direct contact with the gelatin "cell."*

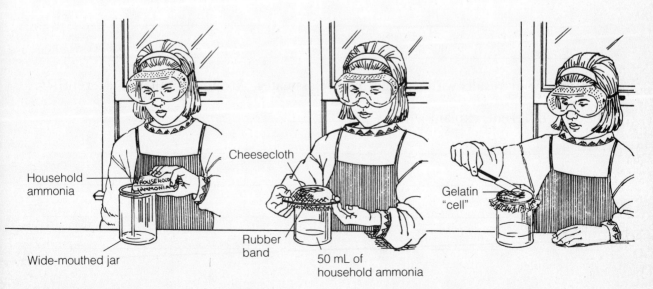

Household ammonia

Wide-mouthed jar

Cheesecloth

Rubber band

50 mL of household ammonia

Gelatin "cell"

5. Note the color of the gelatin "cell" immediately after placing it on the cheesecloth. Record its color in the Data Table.

6. Observe the gelatin "cell" every 2 minutes for a total of 10 minutes. Record your observations in the Data Table.

Observations

DATA TABLE

Time (min)	Color Change
0	
2	
4	
6	
8	
10	

Analysis and Conclusions

1. Does the ammonia diffuse into the gelatin "cell" or does the material in the gelatin "cell" diffuse into the ammonia in the jar? How do you know? _____

_____

_____

_____

_____

2. What causes the gelatin "cell" to undergo changes without coming into direct contact with the ammonia? _____

_____

_____

3. Compare your results with those of your classmates. Are they similar? Different? If they are different, explain why. _____

_____

_____

_____

_____

Activity

Across a Crowded Cell?

Just as you may have a tendency to move from a crowded room to a less crowded room, so too do most molecules. They move from areas of higher concentration (crowded) of that substance to areas of lower concentration (less crowded) of that substance. This process is called diffusion. By doing this activity, you will see diffusion occurring right before your eyes.

Materials

scissors	24-cm length of glass tubing (3 cm in diameter)
strip of red litmus paper	glass stirring rod
filter paper	dilute household ammonia
2 corks (3 mm in diameter)	timer with second indicator
straight pin (less than 2.5 cm in length)	2 medicine droppers

Procedure 🜂 ⬚ ⬛

1. Select one member of the group for each of the following roles: Principal Investigator, Timer, and Observer/Recorder. Make sure you understand your role in the activity before you continue.

2. Using scissors, cut a strip of red litmus paper into six equal pieces. Then cut a dime-sized circle from a piece of filter paper.

3. Attach the circle of filter paper to a cork with a straight pin as shown in the diagram. Make sure that the straight pin does not extend beyond the length of the cork.

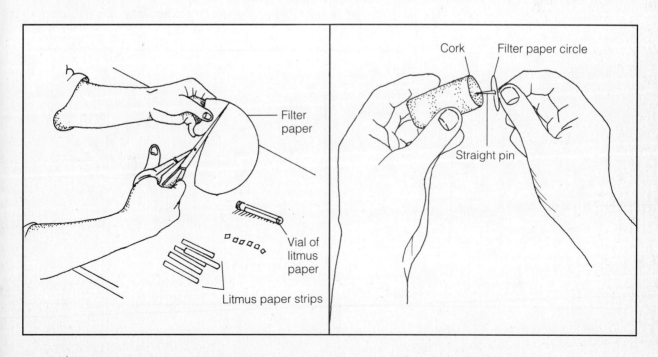

Filter paper

Vial of litmus paper

Litmus paper strips

Cork   Filter paper circle

Straight pin

4. Moisten the pieces of litmus paper with water.

5. Using a glass stirring rod, push the pieces of red litmus into the glass tubing in such a way that they are equally spaced. See the diagram below. The pieces of red litmus paper will stick to the inside of the glass because they are wet.

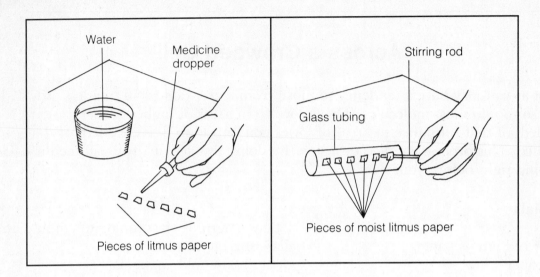

6. Insert the second cork into one end of the glass tubing.

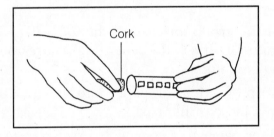

7. Place 2 drops of ammonia on the filter-paper circle that is attached to the cork. Quickly insert the cork into the other end of the glass tubing. **CAUTION:** *Keep the ammonia away from your skin and do not inhale its vapors. Keep the room well ventilated.*

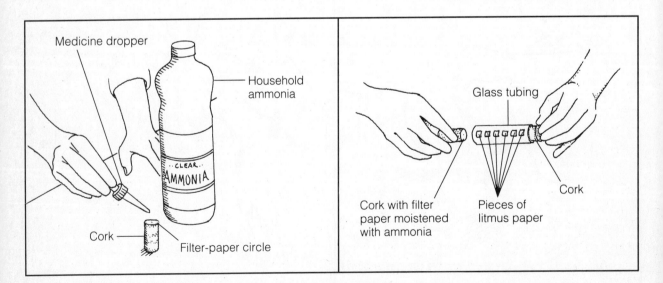

8. Note the time it takes each piece of red litmus paper to turn completely blue. Ammonia is a base and will turn red litmus blue. Record your observations in the Data Table.

Observations

DATA TABLE

Piece of Litmus Paper	Time It Took to Change to Blue (min)
1	
2	
3	
4	
5	
6	

Analysis and Conclusions

1. What process does this activity illustrate? _____

_____

2. How can you tell if this process has occurred? _____

_____

_____

_____

3. Which part of the glass tubing had the highest concentration of ammonia?

_____

The lowest? _____

4. What relationship does this activity have with the process of diffusion?

_____

_____

_____

Going Further

Repeat this activity using a much stronger solution of ammonia. How do the rates of diffusion compare?

Activity

Seeing Is Believing

Not only does a green plant need sunlight and chlorophyll to perform photosynthesis, it also requires carbon dioxide and water. How does carbon dioxide, which is a gas found in the atmosphere, get inside the plant? The answer: It enters through tiny openings on the surface of the leaf. These openings are called stomata (singular: stoma). To see stomata up close, try this activity.

What You Will Need

medicine dropper
microscope slide
lettuce leaf
forceps
coverslip
microscope

What You Will Do 🧪

1. Place a drop of water in the center of a slide.

2. Carefully bend and tear a lettuce leaf toward the main vein as shown in the diagram. Then with forceps pull off strips of the thin membrane on the underside of the leaf.

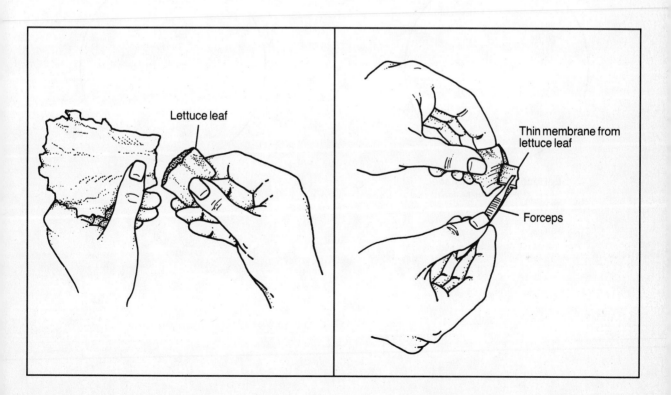

Lettuce leaf

Thin membrane from lettuce leaf

Forceps

3. Place the thin membrane on the drop of water on the slide. Make sure that the membrane is flat.

4. Place a coverslip over the drop of water and leaf membrane.

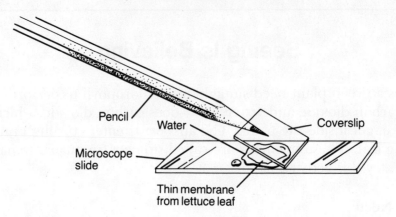

5. Put the slide on the stage of a microscope. Using the low-power objective, locate a few stomata. You may have to adjust the coarse-adjustment knob until the stomata come into focus. Draw and label what you see in the appropriate place in What You Will See.

6. Switch to the high-power objective and locate one stoma. Draw and label what you see in the appropriate place in What You Will See.

What You Will See

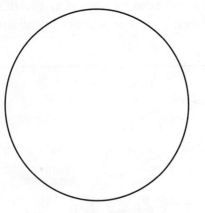

Lettuce Leaf Membrane
(Low power)

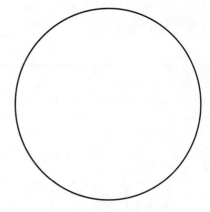

Lettuce Leaf Membrane
(High power)

What You Will Discover

1. In addition to carbon dioxide, what other substance passes through the stomata?

   _____

   _____

2. On each side of a stoma, there is a sausage-shaped cell called a guard cell. What do you think the function of a guard cell is? _____

   _____

   _____

   _____

3. What would happen to a green plant if its stomata were clogged? _____

   _____

   _____

   _____

   _____

   _____

Going Further

   Get together with your classmates and design an experiment to see if your answers to questions 1 and 3 in What You Will Discover are correct. With your teacher's permission perform your experiment. Be sure to state your hypothesis and to include a control.

Activity

Cell Energy

CHAPTER
4

Lighten Up

In order to carry out photosynthesis, a green plant must have light. But how much light? To show how different intensities (amounts) of light affect photosynthesis, why not do this activity.

Materials

test tube
400-mL beaker
freshly cut *Elodea* sprig
forceps
bright light
sodium bicarbonate solution
hand lens
clock with second indicator

Procedure 🔥 🧰

1. Select one member of the group for each of the following roles: Principal Investigator, Timer, Counter, and Observer/Recorder. Make sure you understand your role in the activity before you continue.

2. Completely fill a test tube and a beaker with a sodium bicarbonate solution. Sodium bicarbonate will provide a source of carbon dioxide.

3. Using forceps, place an *Elodea* sprig about halfway down in the test tube. Be sure that the cut end of the plant points downward in the test tube.

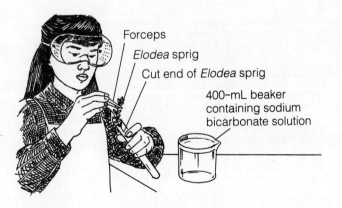

Forceps
Elodea sprig
Cut end of *Elodea* sprig
400–mL beaker containing sodium bicarbonate solution

4. Cover the mouth of the test tube with your thumb and turn the test tube upside down. Try not to trap any air bubbles in the test tube.

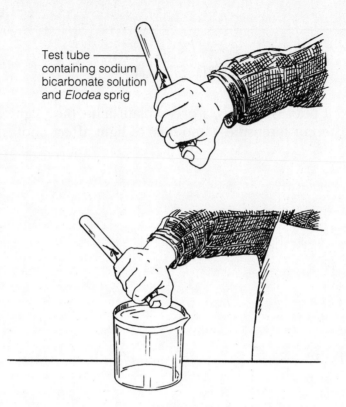

Test tube containing sodium bicarbonate solution and *Elodea* sprig

5. Place the mouth of the test tube under the surface of the sodium bicarbonate solution in the beaker. Remove your thumb from the mouth of the test tube.

6. Gently lower the test tube inside the beaker so that the test tube leans against the side of the beaker.

7. Put the beaker in a place where it will receive normal room light. Using a hand lens, count the number of bubbles produced by the *Elodea* in the test tube for 5 minutes. Record the number of bubbles in the Data Table.

8. Turn down the lights in the room and count the number of bubbles again for 5 minutes. Record the number in the Data Table.

9. Turn up the lights in the room and shine a bright light on the *Elodea*. Count the number of bubbles produced in 5 minutes. Record the number in the Data Table.

Observations

DATA TABLE

Light Intensity	Number of Bubbles Produced in 5 Minutes
Room light	
Dim light	
Bright light	

Analysis and Conclusions

1. From what part of the *Elodea* did the bubbles emerge? _____

_____

_____

_____

2. How does counting bubbles measure the rate of photosynthesis? _____

_____

_____

3. When was the greatest number of bubbles produced? _____

_____

The least? _____

_____

_____

4. How does the intensity of light affect the rate of photosynthesis? _____

_____

_____

_____

_____

5. How do your results compare with those of your classmates? Are they similar? Different? How can you account for any differences in the numbers of bubbles produced? Can you identify any trends even if the actual numbers differ?

_____

_____

_____

_____

_____

_____

Going Further

Perform the activity again using different colors of light. What effect does each color have on the rate of photosynthesis?

— **A**ctivity _____ Cell Energy

CHAPTER

4

Yeasts in Action

Unlike most organisms, yeasts (single-celled fungi) obtain their energy through fermentation. During fermentation, sugars and starches are broken down into alcohol and carbon dioxide. At the same time, energy is produced. To see the process of fermentation in action, try this activity.

Materials

balance
deep jar
sugar
baker's yeast
bromthymol blue solution
graduated cylinder
one-hole rubber stopper
30-cm glass tube with a bend at both ends

Procedure 🜃 ⚕

1. Place 240 mL of water in a deep jar. Add about 15 grams of sugar to the water. Stir the solution.

2. Add a pinch of baker's yeast to the sugar solution. The mixture of water, sugar, and yeast is now called a yeast culture.

3. Pour 100 mL of bromthymol blue solution into a graduated cylinder.

Baker's yeast

Jar

Sugar solution

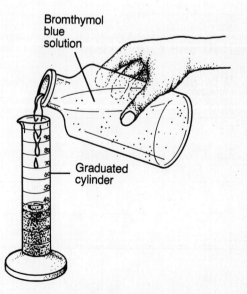

Bromthymol blue solution

Graduated cylinder

4. Insert a one-hole rubber stopper into the jar.

5. Insert one end of the glass tube through the opening in the stopper and the other end in the bromthymol blue solution as shown in the diagram. The bromthymol blue solution will turn yellow if carbon dioxide is present.

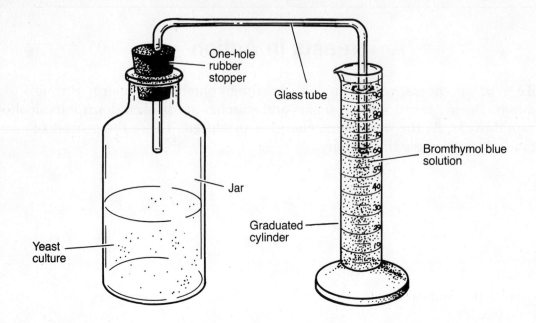

6. Place the apparatus in an area where it will remain undisturbed for at least 1 hour.

7. After 1 hour, observe the yeast culture and the bromthymol blue solution.

Observations

1. Describe the odor of the yeast culture after 1 hour. _____

_____

_____

2. What was the color of the bromthymol blue solution after 1 hour?

_____

Analysis and Conclusions

1. Why did the bromthymol blue solution change color? _____

_____

_____

_____

_____

_____

2. What does this activity tell you about how yeasts get their energy? _____ __

_____

_____

_____

_____

3. Is fermentation anaerobic (does not need oxygen) or aerobic (needs oxygen)?

_____

4. Write a word equation for fermentation using the reactants and products in the

activity. _____

_____

_____

Cells: Building Blocks of Life   D ■ 121

Answer Key

Activity: Hydra Doing?

Observations. 1. The hydra draws in its tentacles. **2.** The hydra moves by somersaulting, or cartwheeling. **3.** The hydra captures food with its tentacles.
Analysis and Conclusions 1. Able to move, respond to a stimulus, reproduce, take in food, and is made of cells. **2.** The hydra reacts as a whole organism because it has no specific pathway for the carrying of incoming messages as humans do (nervous systems). Instead, the messages gradually spread throughout the hydra's body.

Activity: What Is It?

DATA TABLE 1 White; blue-black; yes.
DATA TABLE 2 Golden brown; orange; yes.
DATA TABLE 3 Colorless; purple; yes.
DATA TABLE 4 Yes; yes. **Analysis and Conclusions 1.** Potato; honey solution.
2. Egg-white mixture. **3.** Vegetable oil.
4. Some food inside the bag contains fat.

Activity: Now You See It—Now You Don't

Observations 1. The chloroplasts are evenly distributed throughout the cell. **2.** The chloroplasts were bunched together in the center of the cell. **Analysis and Conclusions 1.** Water moves out of the cell.
2. The cell membrane and the contents of the cell shrink away from the cell wall. **3.** The cell membrane and cell contents return to their original positions in the cell.

Activity: Coming and Going

Teacher Note: Prepare the gelatin "cells" the day before class. Mix 2 packages of gelatin together according to the directions on their packages. Before pouring the gelatin into a clean, empty quart milk carton, add 10 mL of phenolphthalein and stir. Allow the gelatin to cool and set in the carton over night. Before class, lay the carton horizontally on a level cutting surface. With a pencil, mark off as many cross sections of gelatin as you have students or student groups. With a sharp knife, cut through the carton at the marks. Remove the pieces of carton from around the gelatin "cells." **Observations** The "cell" begins to change to increasingly deeper shades of pink. **Analysis and Conclusions 1.** The fact that the gelatin "cell" turns pink and the ammonia undergoes no changes indicates that the ammonia diffuses into the "cell." **2.** The ammonia liquid forms ammonia gas (students should be able to smell ammonia's aroma) that diffuses through the air in the beaker to the gelatin "cell."
3. Class results should be similar to individual student results.

Activity: Across a Crowded Cell?

Observations The farther away the piece of red litmus paper is from the ammonia, the longer it will take to change color. **Analysis and Conclusions 1.** Diffusion. **2.** The pieces of red litmus paper change color from red to blue, indicating the presence of a base—ammonia. **3.** The part of the glass tubing nearest the filter-paper circle soaked with ammonia; the end opposite the filter-paper circle. **4.** The activity shows how ammonia moves from a area of higher concentration (filter-paper circle soaked with ammonia) to an area of lower concentration (pieces of litmus paper).

Activity: Seeing Is Believing

What You Will Discover 1. Oxygen. **2.** To regulate the passage of material into and out of the stoma. **3.** The plant would die because carbon dioxide, which is needed for photosynthesis, would not be able to enter the plant through the clogged stomata.

Activity: Lighten Up

Teacher Note: To prepare the sodium bicarbonate solution, mix 0.5 g sodium bicarbonate with 99.5 mL water. You may wish to have students practice placing a test tube filled with water into the beaker before using the sodium bicarbonate solution and *Elodea* sprig. Students should count only the bubbles coming from the *Elodea* sprig. The bubbles of air that may form on the side of the beaker should be disregarded. The bubbles should start to escape from the *Elodea* sprig about 5 minutes after being placed in bright light.   **Analysis and Conclusions 1.** From the cut stem.   **2.** The rate at which oxygen is produced is a direct measure of the rate of photosynthesis.   **3.** The greatest number of bubbles was produced in bright light; the least number of bubbles was produced in dim light.   **4.** As the intensity of light increases, the rate of photosynthesis

increases.   **5.** Student results should be similar.

Activity: Yeasts in Action

Observations 1. It has the odor of alcohol. **2.** The bromthymol blue solution changes color from blue to yellow.   **Analysis and Conclusions 1.** The bromthymol blue solution changes color from blue to yellow because carbon dioxide was present. **2.** Yeasts get their energy by breaking down the sugar in the sugar solution into alcohol and carbon dioxide.   **3.** Fermentation is anaerobic.   **4.** Sugar → Alcohol + Carbon dioxide.